Daniel Lautenbacher

20 Minutes for good Hardware Knowledge in Personal Computer Systems

GRIN Verlag

Bibliografische Information der Deutschen Nationalbibliothek:

Die Deutsche Bibliothek verzeichnet diese Publikation in der Deutschen National-
bibliografie; detaillierte bibliografische Daten sind im Internet über http://dnb.d-
nb.de/ abrufbar.

Imprint:

Copyright © 2012 GRIN Verlag GmbH
Druck und Bindung: Books on Demand GmbH, Norderstedt Germany
ISBN: 978-3-656-11339-3

This book at GRIN:

http://www.grin.com/en/e-book/187681/20-minutes-for-good-hardware-knowledge-
in-personal-computer-systems

Daniel Lautenbacher

20 Minutes for good

Hardware Knowledge

in Personal Computer Systems

German Original Version
Title: 20 Minuten für gutes Hardwareverständnis bei Personal Computersystemen
ISBN: 978-3-640-90731-1
Year: 2011 | Publishing Company: GRIN Verlag GmbH, Munich, Germany

Index

Preface

In this short book, I would like to introduce you in the construction of a personal computer in a short time. For this, I will show you in the progress of the book the various components of a personal computer.

The aim of this book is to provide you with the knowledge about these components in more detail by learning:

- *which functions the individual components perform.*

- *which properties these components have.*

- *in which way the single components fit together.*

- *how you will be able to affect the stability and speed of a system.*

- *how you build a PC-System by yourself based on this knowledge.*

In this book I will point out only the necessary details to accomplish these targets.

Following this, you will find a checklist at the end of the book.

 ## Important Notice:

Always disconnect the computer and all components of the hardware from the Power Supply before you change any components or when you are (re)assembling the components or other parts!

Make sure that your body has no electrostatic charge. To ensure this, please touch various metallic bright surfaces such as radiators to discharge. An ESD shock can destroy your hardware!

Always follow the instructions and warnings that are listed in the manuals of the individual components.

The author assumes no liability for damages of any kind, whether for personal injury that may result from improper or incorrect use of this book. This book is for information purposes only and is not an assembly manual. For errors, mistakes and misprints the author assumes no liability. Claims for damages for any reason are excluded. If any provision is invalid, the remaining provisions shall remain unaffected.

Product names, trademarks and registered trademarks are the property of their respective owners and are not usually marked as such. The use is for information only. The absence of such indications does not mean that it is a free name in terms of trademark and trademark law. The author recognizes all product names and trademarks. All rights reserved. No part of this publication may be reproduced in any form without written permission of the author or electronically processed, copied or distributed.

Basic components of a computer.

THE MAINBOARD, also called Motherboard, is the beating heart of a Personal Computer System. You will find all important parts of a computer on this. The Mainboard is connected with the Power Supply to power the main components.

This includes:

- Processor (CPU)
- Memory
- Video Card

When you select a motherboard you must pay attention to the Form Factor (the size of the motherboard), the chipset, the type and speed of the RAM (memory) mountable, and furthermore to the socket ("port of the processor").

The two most important Form Factors are ATX and Micro-ATX(µATX).
ATX Mainboards are bigger than Micro-ATX Mainboards, so they will not fit in every usual chassis. You should ensure that the board you have chosen fits into the case.

THE CHIPSET contains the most important functions of a motherboard:
The type and speed of the Memory, Processor, USB, SATA, PCIe ports etc. depend on the chipset in type, quantity and speed.

The type and speed to find a suitable memory module can't be taken from a usual trade term. Because of this we will cover this topic in the category Memory.

ON THE SOCKET the Processor is mounted. There are many different types of sockets, including the AM3 socket from AMD and the Intel socket 1366.

A usual trade term for a motherboard is:
Asus M4A79T Deluxe 790FX AM3 ATX

Now we split this term into the several parts:

Manufacturer	Model	Chipset	Socket	Form Factor
Asus	M4A79T Deluxe	790FX	AM3	ATX

You see, you will perceive the most important information at first glance.

If you buy a Motherboard, pay attention to the suitable Power Supply. Electricity connectors are recorded as follows: „24 Pin + 8 Pin" or „ATX 2.x" or „EPS" etc.

THE PROCESSOR is the central processing unit (CPU) of a computer system. He takes over almost all calculations in a computer needed to run a program. The speed (frequency) of a processor is measured in GHz. It has its own "small memory" which is divided into different layers (level 1, level 2 and level 3 cache). In these cache it is possible to outsource major parts of a program.

The two major manufacturers are Intel and AMD.

Processors use different sockets, as we have noted in the previous part. Below is a table in which the major sockets could be associated with their vendors:

AMD	Intel
AM2	775
AM2+	1155
AM3	1156
AM3+	1366

As you can see, you can recognize the manufacturer very easily if you take a look at the socket.

Most modern processors have multiple cores, which enables them to execute several arithmetic operations **simultaneously.**

Processors are usually sold in two versions, boxed and tray.
In the boxed version you get a stock cooler for the processor in addition. With the tray variant you will only receive the processor.

Important: A processor must **always** be cooled!

When you are buying a cooler, you have to pay attention to the socket, and you have to use the included thermal paste. This paste must be applied between the cooler and the processor.

A commercial designation for an AMD processor could be for example:

Vendor	Model	Cores	Performance	Speed	Socket	Version
AMD	Phenom II	X6	1055T	6x 2.80 GHz	So.AM3	Box

With Intel Processors you should know that not all cores are real physical cores. Intel uses the HT-Technology which allows it to use virtual cores.

Vendor	Model	Cores	Performance	Speed	Socket	Version
Intel	Core	i7	950	4x 3.06 GHz	So.1366	Box

THE RANDOM ACCESS MEMORY (RAM) is another major component. In this component, large parts of programs and data currently in use are cached. Because the memory reaches a much higher read speed than hard disks, this allows better performance. If the memory is too small, parts still need to be repeatedly loaded from the hard disk before they can be processed. This can result in massive speed dips.

At the present time (January 2012) I would recommend the use of 4 GB of RAM up to 8 GB if you use a 64-bit (x64) operating system (e.g. Windows 7 is available in 32- and 64-bit versions).

Make sure that the modules are always from the same manufacturer (for example, 2 x 2 GB). A motherboard usually has 4 slots, each for a memory module, these 4 slots are further divided into two colors (eg red and blue).

I recommend the use of 2 or 4 modules of the same size, speed and the same manufacturer. If you are using only 2 modules, make sure that the modules are located in the two slots of the same color. Insert here the first module in the slot that is closest to the processor. So you achieve the best possible result.

Memory is currently available in two variants who are important to us.
(Type, in the course "Ram-Type" called):

DDR2-Ram
DDR3-Ram

Notice that SO or SO-Dimm Memory is used for Notebooks.

Your motherboard and processor usually support only one of these memory types, based on your chipset and memory controller or the processor. To help you, please notice the following table:

Socket	RAM-Type
AM2	DDR2
AM2+	DDR2
AM3	DDR2 / DDR3*
AM3+	DDR3
775	DDR2 / DDR3*

Socket	RAM-Type
1155	DDR3
1156	-*
1366	DDR3
*Depends on the used processor.	

Take a look on the details of your mainboard to identify the maximum speed and type of RAM supported by your board.
The higher the clock rate and the lower the latency, the faster the RAM.

Commercial designation:

Size	Vendor	RAM-Type	Speed/Clock r.	Modul	Latency	Quantity*
4 GB	Mushkin	DDR3	1600	DIMM	CL7	Dual Kit*

* in this case 2 RAM modules (thus 2 x 2 GB).

THE GRAPHICS CARD is responsible for the calculation of images in the computer. Modern graphics cards use the PCIe (PCI Express) connector on the motherboard. They are inserted into the slot closest to the processor, usually at the top. On some motherboards, this port is also highlighted.

Make sure that your motherboard has such a connection, with the speed of x16, usually indicated by "PCIe x16".

Some modern graphics cards also require additional power from the power supply of the computer, paying particular attention to the details of your graphics card and power supply, the necessary power connections are usually recorded as follows: "1x PCI-Express 6 +2- pin" / "1x PCI-Express 6-pin"

For graphics cards, you still have to note some additional factors such as:

- **Memoryinterface / Memorybus:** I recommend the use of graphic cards with a minimum of 256 bit memory interface.

- **GPU-Clock:** The clock rate of the graphics processor.

- **GRAM-Clock:** The Clock rate of the graphics memory (the memory of the graphic card).

- **GDDR:** The type of the graphics memory.

- The size of the graphics memory

The higher the clock of the GPU and VGA memory, the faster the graphics card, assuming the memory interface is not limiting it.

The two major chip manufacturers are ATI / AMD (Radeon HD) and NVIDIA (GeForce).

If you use CAD applications you have to buy a workstation graphics card.

If you want to use your computer for your home or for movies and games, you should **not** buy a workstation graphics card.

Commercial designation for AMD/ATI – Graphic Cards:

RAM-Size	Vendor	Chip-Type	Performance ratio	Type	Connection
4096 MB	Sapphire	Radeon HD	6990	GDDR5	PCIe

THE HARD DRIVE is the physical disk of a computer. On this drive, all data is permanently stored and are still available even after a reboot.

In modern hard drives there is not much to note. Most hard drives use a SATA connector and require a power supply with a SATA power connector.

The hard drive is connected to the motherboard via a SATA cable.

One should note: The size, rotation speed, the form factor and the cache.

A modern hard drive should have 500 - 1000 GB and a rotational speed of 7200 rpm. A disk with a lower rotation speed is slightly slower, but (in most cases) power saving and low noise, the latter should be the main reason for a slower rotating hard disk.

The cache of a hard disk is again something like a "small memory", just at your hard drive. Usually a hard drive has a cache of, 8MB, 16MB, 32MB or even 64MB. Although I personally always prefer the largest possible cache. This is also useful if you are working with very large files.

The form factor of a hard drive gives the size of the hard drive, essentially there are two sizes:
3,5 Inch or 3.5 " – 3,5-Inch-Hard Drives are "normal Hard Drives"".
2,5 Inch or 2.5 „ – 2,5-Inch-Hard Drives are normally used in notebooks because they are smaller than the normal ones.

Commercial designation:

Size	Vendor	Model	Cache	Form factor
2000 GB	Western Digital	Caviar Green WD20EARS	64 MB	3.5"

OPTICAL DRIVES usually also have a SATA connection and need a SATA power connector. Internal optical drives usually have a size of 5.25 inches.

Optical drives are required for burning and reading CDs, DVDs, Blu-Ray-Disks, etc. Make sure that you buy a drive that you want to / can actually use. Note that burning drives also are able to read disks.
Bulk = Without Tools | Retail = any colored box + perhaps manual, software

Commercial designation:

Vendor	Type	Model	Connector	more
LG Electronics	DVD-Brenner	GH22NS50	SATA	Black Bulk

THE POWER SUPPLY provides all your components with power. When choosing a power supply, you should pay attention to the existing connections and also to the quality and performance information. You must ensure that the power supply can provide all the existing components with power, both on the number of connections, as well as the amperage and the wattage.

In the following example we will use the information from the
"Be Quiet! Straight Power E8 500 watt" power supply.

Is the main connector you need for your motherboard mentioned on the list?

1 x ATX (20-pole), cable length 55 cm or
1 x ATX 2.x (24-pole), cable length 55 cm or
1 x EPS (24-pole), cable length 55 cm

Some motherboards will also need an additional port, is it also listed?

1 x ATX12V (8-pin), 1x 70 cm cable length

Are there enough ports for your drives?

5 x 5.25 inches
2 x 3.5 inches
6 x SATA

Are the right connections available for your graphics card?

1x PCI-Express 6 +2- pin
1x PCI-Express 6-pin

Have you chosen the right design?
ATX

Make sure the current strengths of individual lines. Feel free to use the following information as a guide. Note however that some power supplies have only two 12 V rails, pay attention particular the fact that each line has at least 21 amps.

+3.3 V	24 Amp
+5 Vsb	3 amps
+5 V	22 Amp
+12 V1	18 amps
+12 V2	18 amps
+12 V3	18 amps
+12 V4	18 amps
+12 V	total: 36 amps
-12V	0.3 Amps

The higher the amperage 12V lines, the more the better.

In addition to the current course, the wattage also plays an important role. But don't let yourself be fooled by the specified watts! Some power supplies with 1000 watts only offer the same performance as a 500-watt power supply.

To find out how much watts should your power supply have:

- Calculate for your motherboard 50 watts, 100 watts, if your motherboard has an onboard graphics card.

- Calculate for each hard drive 10 Watt.

- For each RAM-Module 2 Watt

- For each cooler 1 Watt

- For each optical drive 20 Watt, for burning drives 30 Watt.

- How many watts at full utilization of your processor and graphics card are used, you could read in the details. This value is advertised as TDP, for example, "TDP: 45 Watts."

Also, it does not hurt to buy a PSU (power supply unit) with special protection functions such as protection against spikes (OCP), over voltage protection (OVP), under voltage protection (UVP), overload protection selected (OLP / OPP), Short circuit protection (SCP) and thermal overload protection (OTP).

If you want to save energy, you have to pay attention to the efficiency of your power adapter and choose a power supply with an efficiency of 80 percent or more.

Here are some examples of commercial designations:

Watts	Vendor	Model	Efficiency	Type
550W	Be quiet!	System Power BQT S6	80+	BULK

Watts	Vendor	Model	Efficiency
650W	XFX	Pro Core Edition	80+ Bronze

Bulk=Without Extras | Retail = any colored box + perhaps manual, software

THE CASE is, as the name suggests, the appearance of your computer. There are different housing types, and not every motherboard will fit into every case.

Therefore, make sure that:

- the design of your motherboard is listed within the details of your case (eg ATX).

- you have enough space for your hard drives (3.5 inch bays).

- you have enough space for your other drives (5.25 inch bays).

There are three types of cases that are important for us:

- Mid-Tower, which are "normal" large cases, usually placed vertically on the floor. A midi-tower is about 43 cm in height.

- Big-Tower are slightly larger cases, which will also be placed vertically on the floor. A big tower is about 60 cm high.

- Desktop cases are basically the same size as Midi-Tower, the difference is that they are mostly flat, e.g. put on a table. A desktop housing is approximately 39 cm wide, 12 cm high and 41 cm deep.

Which case you choose is up to you. However, as already mentioned, please ensure that the case you want to use provides enough space for all components you need.

Some cases are even sold with a pre-built power supply. I would advise you not to buy such a case.

Commercial designation:

Motherboard Size	Vendor	Model	Type	NT*	more
ATX	Antec	Gamer Case Three Hundred	Midi Tower	o.NT	Schwarz

NT=Power Supply

Modular construction of a computer

**WHAT DO YOU HAVE TO PAY ATTENTION TO
WHEN YOU COMPOSE A NEW COMPUTER?**

As we have already explained in the previous sections, it is important that all components fit together. Here the design (eg ATX), the base (eg AM 3 +) and the type of memory (eg DDR 3) are the most important parts.

If you have composed the fitting parts, you only need to assemble all components and connect them with each other. It is important to mention that this should be possible without the use of physical force in most cases.

Also note that the processor is not simply put in, you must first move the socket with a small lever. Take a look into your motherboard manual to learn how to install your CPU properly.

The power-button of your case usually is connected to the motherboard with the cable "Power S / W". Read again your motherboard manual for further instructions.

THE SYSTEM-TYPE is an important issue. Do you want to build a computer for your office, for HD movies or playing computer games is also an important question.

- If you want to use your computer only for the office or the work with Word, usually an onboard graphic card is enough.

- If you want to use your computer for watching high-definition movies, a weak graphics card is enough, which works even without additional power supply. As an example here, I would use the "ATI Radeon HD 5450".

- If you use your computer for games, pay attention to the memory interface. Use a 256-bit graphics card and make sure that you have at least 4 GB of system memory.

Watch out also that the components fit together not only technically. In a system with an AMD processor, it is obvious that using a motherboard with an AMD chipset and AMD / ATI graphics card might be more stable than any other combination of components.

This could of course also correspond to a combination consisting of an NVIDIA graphics card with an NVIDIA motherboard chipset.

Important Interfaces

PCI (Peripheral Component Interconnect) is an interface that allows other components to connect the processor with the chipset in the form of plug-in cards to provide additional ports or functionality.

PCIe (Peripheral Component Interconnect Express) is the successor to PCI, which allows (in comparison to his predecessor) higher data transfer rates. This interface allows it to incorporate modern graphics cards (among others) in computers.

SATA (Serial ATA) is intended for the exchange of data between the hard disk and the processor. The connection is via a ribbon cable between hard disk and motherboard.

USB (Universal Serial Bus) is an interface for the universal connection of external devices to the computer with only one port. The particular advantage of this interface is that it is possible to connect an external device even while the system is running.

FireWire (i.Link / IEEE 194) is an interface developed by Apple for the rapid exchange of data between the computer and external devices.

DVI (Digital Visual Interface) is an interface for transmitting video data. This port is now most used in connecting the graphics card to the monitor.

HDMI (High Definition Multimedia Interface) is an interface for transmitting audio and video data.

Checklist

☐ The CPU Socket fits to the Mainboard

☐ The System Memory fits to the Mainboard/Processor.

☐ All Memory Modules are from the same manufacturer.

☐ The memory modules equally fast.

☐ There is an equal number of RAM modules, 2x2GB or 4x2GB for example.

☐ The Mainboard fits into the case.

☐ The power supply has the correct form factor for the case.

☐ The power supply fits to the motherboard and has the necessary connections.

☐ The graphics card is compatible to the motherboard.

☐ The power supply matches the graphics card and has the necessary connections.

☐ The hard disks and drives fit into the case.

☐ The power supply comes with connectivity options for all disks / drives.

☐ A CPU cooler is available.

Office-Computers:

☐ An onboard graphics card is available.